581
P

c.2

Prince, Jack Harvey.

Plants that eat
animals

DATE			

642

Plants That Eat Animals

Plants That Eat Animals

by
J. H. Prince

ELSEVIER/NELSON BOOKS
New York

Library of Congress Cataloging in Publication Data

Prince, Jack Harvey.
   Plants that eat animals.

   Bibliography: p. 81
   Includes index.
   SUMMARY: Discusses life processes of several types of carnivorous plants including venus flytraps, bladderworts, and sundews. Includes instructions for cultivating these plants in flower pots.
   1. Insectivorous plants—Juvenile literature.
[1. Insectivorous plants]   I. Title.
QK917.P68   581'.5'3   78–14791
ISBN 0-525-66599-4

Published in the United States by Elsevier/Nelson Books, a division of Elsevier-Dutton Publishing Company, Inc., New York. Published simultaneously in Don Mills, Ontario, by Thomas Nelson and Sons (Canada) Limited.

Printed in the U.S.A.

10 9 8 7 6 5 4 3 2

Contents

Plants That Eat Animals

Introduction

$\mathcal{M}$any students of wildlife find it difficult to raise the same interest in plants as they can in animals; and yet in many ways there are great similarities between the two forms of life. Both have single-celled members; both have species with elaborate vascular systems, and both have asexual and sexual species. The main difference between plant and animal life is that plants have no kind of nervous system; but one can be forgiven for not being so sure of that when studying some of the carnivorous plants.

The idea of flesh-eating plants may leave some people incredulous. They would suppose that such things come only from the pens of science-fiction writers conjuring up hydroponic monsters capable of seeking out and consuming anything around which they can entwine themselves. But of the hundreds of species of carnivorous plants in the world, few entwine themselves around anything, and the majority take only prey that is almost always very small, usually insects or small water creatures. Only a few take small birds or lizards that fall into them.

Such plants are known as *saprophytes,* a word which really refers to any vegetable organism that lives on decayed organic matter (*sapro* means "dead"). Bacteria that live in this way are called *saprophiles*.

Although these plants all have the common feature of trapping and digesting animals, they are found among several widely separated groups of plants, and in widely separated parts of the world. It seems obvious that the carnivorous habit evolved because plants that grow in poor acidic conditions need to supplement their nutrition with additional nitrogen products, for when the poor soils are fed with adequate minerals, the plants thrive well without animal food.

Because so many carnivorous plants are quite unrelated, they demonstrate clearly the way nature produces similar structures and habits when conditions are similar, no matter how widely separated in geography and biologically unconnected these plants may be.

1. Classification of Carnivorous Plants

$\mathcal{A}$ll plants are divided into four types, called *divisions*. These have rather formidable names, but their descriptions are easy to remember, and so for the present these descriptions should be accepted rather than any effort made to memorize the names. We find carnivorous plants in only two of the four divisions, the first and the fourth.

The first division consists of the Thallophyta, or plants without roots, stems, and leaves, in which the plant body is called a *thallus*. These plants have no vascular (blood-vessel) system. They almost sound like "nothing" plants, but the algae, fungi, and bacteria that make up this division are very important to the balance of life on earth. Of the three forms we shall be considering only the fungi, and it will be seen that many of them have quite sophisticated traps, which is surprising in such lowly and often tiny organisms.

Plants in the second division, the Bryophyta, are moisture- and shade-loving plants confined to dry land; mosses and liverworts. They have hairlike filaments called *rhizoids* for carrying water.

Plants of the third division, the Pteridophyta, have distinct roots, stems, and leaves with well-developed vascular systems that bring water and nourishment to the leaves from the main stems, such as ferns and horsetails. These plants are also called *cryptogams (crypto* means "hidden"). They do not have flowers and bear no seeds, however.

Finally, the fourth division, the Spermatophyta, consists of flowering or seed-bearing plants. This is the main division with which we are concerned here.

The serious student will know that divisions are broken down into subdivisions, which are divided into classes; these are reduced to orders, and the orders to families, genera, and species. But this book will remain relatively simple by dealing only with the essentials from families to species.

Because many plants are called *monocotyledons,* we should examine this word. When applied to botany, a *cotyledon* is a primary leaf, the first leaf developed by a seed, so monocotyledon means having a single first leaf on the seed (*mono* means "one"). A *dicotyledon* (*di* means "two") has a pair of leaves sprouting from the seed. The leaves of monocotyledon plants have parallel veins, whereas those of the dicotyledons have networks of veins very much like animals. The two kinds of plant are therefore very easy to distinguish. Most carnivorous plants are dicotyledons, but monocotyledons are found among them as well.

The carnivorous fungi and flowering plants described in this book have been organized into five main groups:

1. Pitfall traps: hollow vessels holding liquid into which animals fall.
2. Flypaper traps: plants with sticky leaf surfaces that trap flies and other insects and hold them until they are digested.
3. Snap traps: specialized leaves that snap shut and form an enclosing chamber when hair triggers are touched by insects.
4. Suction traps: specialized water plants with bladders that draw in prey much as a fish sucks smaller fish into its mouth.
5. Adhesive traps: ring-trap, spore, and bait-trap fungi.

Scientific names are given in some instances. They may look very formidable, but for most purposes can be ignored until the subject is well understood.

The pitfall traps dealt with in Chapter 3 are pitcher-type plants forming three families, the names of which are given in Table 1 (page 16). Flypaper traps with sticky leaves or sticky globules on fine stalks projecting from the leaves are of three kinds, and these are listed in the same table, with full descriptions in Chapter 4. One genus, *Drosera,* includes more than ninety species which occur all over the world, but particularly in Australia and South Africa, as well as a one-type genus[35] (see Figure 28), which is confined to the sandy soils of southern Spain, Portugal, and Morocco.

The Lentibulariaceae have well over three hundred species, and about thirty of them are flypaper traps. These are found in both the Northern and Southern Hemispheres. As many as possible of these are illustrated and described later, so that the reader need not be concerned with their family names before getting an understanding of their nature and structure.

The fourth group in table 1, the snap traps, are found in one form or another in almost every part of the world. A species found in North and South Carolina[23], which closes over any insect alighting on it, is probably the best known of all the carnivorous plants; see Figures 31, 32, 33. There are underwater genera of snap traps that take insects and small fish[4, 65–71]. These are the most complex kind of plant to be found.

And finally there are predaceous forms of fungus that trap protozoa and worms.

There have been rumors of a bird-eating plant in northeastern Australia, but this is probably a myth. However, there are some very large pitcher plants, and it is possible that a tiny bird fell into one of these when trying to get at the nectar and died there. Its remains, found after decomposition, may have given birth to the idea that these plants eat birds. Although these plants will consume any animal tissue, so far there has been no scientific observation of such an occurrence in that particular area.

There are giant pitcher plants in India and Borneo, however, which are so large that they will, in fact, drown a bird or other small animal, whereas plants that form rosettes of leaves to hold water have been known to trap frogs and lizards. Each of these plants will be described more fully later.

Almost all carnivorous plants grow in bogs, poorly drained soil, or in soil where there is a shortage of nitrogenous compounds. There are many elements essential to plant growth. These include boron, calcium, potassium, magnesium, phosphur, and sulfur, but nitrogen is probably the most important; growth can hardly take place without it. Nitrogen plays a vital part in the structure of proteins, which are the major constituent of protoplasm. But some plants are unable to convert nitrogen for use, and it must be converted by some other agency, something that will break down the remains of living matter into simpler substances and release the nitrogen. Carnivorous plants have become adapted to do this.

When adequate soil nutrients are available, these plants thrive without animal food. In fact, many of them can be raised in suburban gardens, and almost all will live when protected from weather extremes in a greenhouse. How this can be done will be described in Chapter 8. But experiments have shown that in spite of their ability to grow without an animal diet, the plants do yield much more prolific crops of seeds when fed with animal tissue.

What is so remarkable about these carnivorous plants is the way in which they reduce their prey to a solution that can be absorbed through the walls of their leaves. Ordinarily, no plant can digest animal tissue, because digestion must take place inside cells; but in carnivorous plants it occurs outside them through the use of external digestive juices called *proteolytic enzymes,* which break down proteins into amino acids and peptones—simpler compounds that the leaf can absorb. These enzymes, secreted by special glands, will cover the captured prey, reducing its proteins to soluble form. The products of this external digestion are then easily absorbed by the plant.

Glands. Two kinds of glands are found on the surfaces of carnivorous plants. They are *trichomes,* which are outgrowths of the epidermis, or skin membrane, of a plant; and *emergences,* which are more substantial, emerging from some of the deeper tissues. Trichomes may be anything from single-celled hairs to complex multi-celled structures, many of which are glandular. Typical emergences are the thorns on rose plants.

Although almost all carnivorous plants have these specialized glands to aid them in digestion, they are not all the same either in structure or in the composition of their secretions.

Digestion. Food must always be converted into a solution in order to be absorbed into cells, and it is this conversion that we call digestion. The substances to be used as food by any organism are converted into a solution by the chemical process of decomposition plus *hydrolysis,* which means "taking up water." In carnivorous plants, this process converts proteins into water-soluble peptones and amino acids, which then diffuse through the cell walls of the leaves and supply the plant with its needed nitrogen and perhaps other necessary substances too. Let us look at this a little more closely.

Enzymes. The proteolytic enzymes already mentioned are organic *catalysts;* and the one that breaks down proteins is called *proteinase* or *protease.* This breaks down ALL proteins. Similar actions are produced on starches by *amylases,* and on fats by *lipases.* One can say that enzymes are complex organic compounds that speed up chemical reactions in a manner similar to that of the catalysts in chemistry. The whole life process depends on this action.

One of the remarkable things about enzymes is that they are not only specific in their action, but that a very small amount can effect changes to an almost unlimited extent because it is not destroyed; it is used over and over again. In many of the carnivorous plants (but not in all of them), decomposition bacteria assist the enzymes in breaking down the animal tissues, just as they do in the soil or in the air.

Osmosis. Once animal tissues are broken down into soluble substances, they are taken up through the walls of the leaves by *osmosis.* This is a process by which water and solvents pass through semipermeable membranes, sometimes called "differentially permeable membranes." When we say "semipermeable," we mean that only the small molecules of a solvent can pass through its wall, and not those of any dissolved substance with larger molecules. If all molecules could pass through, the membrane would be permeable.

A semipermeable membrane may be compared to a net that allows small fish, like sardines, to escape through its mesh, but which traps larger fish, like herrings. It is because the size of a molecule controls its passage through a membrane that water will pass through in one direction, to where the fluid is denser; but not in the opposite direction, from a denser to a less dense fluid. The larger molecules in the denser fluid block the passage of the water through that way. This means that the denser fluid becomes diluted by the water that passes into it from the less dense fluid. Eventually the fluids on the two sides

TABLE   1

KIND OF TRAP	FAMILY AND GENUS	NUMBER OF SPECIES	WHERE FOUND
Pitfall	***Nepenthaceae***		
	Nepenthes	68	From the eastern tropics to Ceylon and Madagascar
	Sarraceniaceae		
	Heliamphora	5	British Guiana, Venezuela
	Sarracenia	9	Eastern North America, from Labrador to Florida
	Darlingtonia (Chrysamphora)	1	Northern California and southern Oregon
	Cephalotaceae		
	Cephalotus	1	Southwest Australia
Lobster Pot	***Lentibulariaceae***		
	Genlisea	10	West Africa and eastern South American tropics, Guiana and Cuba

of the membrane become equally dense. In this balanced state the two fluids are said to be *isotonic*—they have equal osmotic pressure. It is the protoplasm which acts as the semipermeable membrane, and since the cell sap has osmotic pressure, this facilitates the process with which we are concerned.

Flypaper	*Lentibulariaceae*		
	Pinguicula (butterworts)	30	Northern Hemisphere, Old and New Worlds
	Byblidaceae *Byblis* (rainbow plants)	2	Western Australia
	Droseraceae *Drosophyllum*	1	S. Portugal and Spain, Morocco
	Drosera (sundews)	90	Worldwide
Snap	*Droseraceae* *Dionaea* (Venus's flytrap)	1	North Carolina and northern South Carolina
	Aldrovanda	1	Europe, India, Japan, Africa, and northern Australia
	Lentibulariaceae *Utricularia* (bladderworts)	275	Worldwide
	Biovularia	2	Cuba, eastern South America
	Polypompholyx	2	South and southwest Australia
	Fungi	50 or more	Worldwide

In the osmotic process, movement of anything through a membrane into a cell is called *endosmosis*. Movement out of a cell is called *exosmosis*. In carnivorous plants we are mostly concerned with endosmosis, and so it is obvious that when an insect falls into the fluid in a pitcher, the animal protein must be broken down and considerably diluted before it can pass through to the protoplasm within. It must become *hypotonic*—that is, have a lower osmotic pressure than the surrounding matter.

The proteins, which are so important to a plant, are compounds of carbon, oxygen, hydrogen, and nitrogen. Most of them also contain sulfur and some phosphorus, almost all the elements, in fact, that are essential to plant growth (see page 14). All have very complex molecules in a combination of amino acids. Amino acids, called the building blocks of protein, are organic acids. A detailed description of their composition can be found by reference to a book on organic chemistry.

It is always useful to have a generic table to which rapid reference can be made, and this list of genera of the various carnivorous plants, their families and locations will prove useful while reading the main text. It would seem from this list that there are around five hundred species of carnivorous plants, and a score of species of carnivorous fungi. The numbers recorded by investigators vary considerably, and there may be more of this kind of fungus still undiscovered in the soils of the world.

There are many other plants whose flowers might seem well equipped to consume animal tissues because they trap insects, but there is no evidence that this ever occurs. Carnivorous properties always appear to be confined to leaves; never to flowers. Some of the orchids, of which Figure 1 is an example[48], have excellent fluid bowls that attract and sometimes temporarily trap insects; but it is doubtful that this is for any purpose other than pollination by the insects, which nevertheless may become imprisoned for good and drown.

The arum lily is an example of an insect trap designed to hold insects by providing an unclimbable flower wall with an oily secretion and a pitfall trap that imprisons them. Once pollination has taken place, however, the condition of the flower wall changes, and the insects can escape. The sole object of the entrapment is pollination, not food.

There are other flowers that act not only more truly like insect traps of the pitcher-plant type, but which have even adopted some of the formations of the true pitchers. However, they are flowers, not leaves. Known as *birthworts,* they belong to a family of woody vines and shrubs, the *Aristolochiceae,* of which there are ten genera

1.   Orchids such as the two species of *Paphiopedilum (top and middle)* trap rainwater in a cup-shaped flower. Insects also find their way in, but there is no process by which a flower can digest animal tissues, and they use insects only for pollination. In plants that eat insects, it is only the leaves which become specialized for this purpose. Some flowers do use real traps, however, and *Aristolochia salpinx* from Paraguay *(left bottom)* has downpointing tentacles to trap temporarily insects that enter for nectar. *Aristolochia cathcartii* from Sikkim *(bottom right)* contains quite a considerable amount of insect-attracting secretion and slippery walls.

and about six hundred species. Some of these give off a smell of rotting meat, and this attracts carrion flies, which crawl inside and pollinate the flowers.

There are some plants that have nectar glands around the lip, and down-pointing toothlike tentacles all the way to the base of the inside of the flower. These prevent the escape of any fly that enters, but once pollination has taken place, the tentacles wither, and the insect escapes. These flowers vary from small to very large, but they are in no way carnivorous.

2. Underwater Traps

There are carnivorous plants that spend their lives under water, except for a period in spring when they must be pollinated. Some of them capture their prey in traps that snap closed on them, and others draw in their prey by suction. These are all found in the swamps and fresh waters of the world, and they have exceedingly complex and delicate mechanisms that enable them to operate with the speed and precision of electronic instruments.

Bladderworts. This is the common name of a number of rootless underwater plants that have small chambers or bladders on stems and leaves. The sides of the bladders are flattened together when they are below the water surface, so they are free of air or water, and at this time they are set to trap small water creatures. But in spring, when they rise to the surface, the bladders fill with air and act only as floats, allowing the plant to flower in the air, where it can be pollinated by insects like any land plant.

The plant will float to some extent even if the bladders are removed, but they do provide more buoyancy. While the plant is floating, it has no leaves; it merely blooms, and as soon as the blooms have been pollinated, the bladders fill with water, the whole plant sinks down once more, its leaves form, and its seeds ripen at the bottom of the pool or lake. The flower is yellow, stands well clear of the water surface, and is a good indication of where the plant may be found.

Bladderworts are found in both tropical and temperate zones of the world. In fact, they have a worldwide distribution. There are forty species in Australia alone, and altogether there are probably nearly three hundred species. One of them[71], *Utricularia vulgaris,* which is found in Europe and America, can be cultivated in garden ponds. Its stems grow from six to eighteen inches in length.

Utricularia is not only the commonest of the bladderworts, but its underwater pear-shaped bladders, of which there is an immense variety, are structurally the most complex, and are said by some experts to be the most advanced form of trap found in the entire plant kingdom.

Some species of *Utricularia* are almost land plants. Their bladders are on runners, and they lie on mossy banks or on the drier edges of swamps. One can imagine them as being a kind of halfway plant between the true underwater traps and those that live only on land. They often have miniature but conspicuous flowers resembling snapdragons, which, in the case of one species [66] called *Utricularia dichotoma,* are commonly called "fairy aprons." They may be shades of purple, gold, or blue.

But most species of *Utricularia* are found under water. They have very unusual leaf constructions, and this includes their bladder traps. The translucent green bladders are often only 2.5 millimeters (a tenth of an inch) in length, but in some species they grow to twice this size. They have no difficulty in trapping daphnia, mosquito larvae, tiny worms, or small fish, and up to ten small crustaceans may be found in a single bladder at one time. Occasionally a trap will snare a tiny tadpole, and if this is too big to engulf, it is held by the head or tail until it dies. It will stay jammed in the door, however, until another creature touches the tripping mechanism and springs the door open again, at which time the dead remnants are drawn in.

The inability of a plant to discriminate the size of prey it traps, so that it snaps up excessively large creatures, eventually results in the death of the bladders in which the too-large prey becomes jammed. This death is considered by some experts to be due only to overfeeding. In any case a bladder usually dies in ten to fourteen days, so its life is never long.

Typical traps have valves or doors and a tripping mechanism with a spring action that opens and closes them, all within less than a thousandth of a second. This action is aided by the hydrodynamic force of water on them as it tries to force its way into the bladders. There are many forms of these traps, and in the commonest there are two valves. The upper valve is the larger, and is really the main door. It is attached with an overhead hinge, and to the side wall as well. There is a stop to prevent its operation without being triggered by prey; and a thin membrane, or velum (the second, smaller valve) protects and seals its edges.

The valves, or doors, carry mucilage glands on stalks, and these secrete a substance that is attractive to daphnia, cyclops, and other small water creatures. Water is pumped out of a bladder by osmosis, and some species are considered capable of expelling almost 90 percent of their water contents in this way. The somewhat concave walls of a bladder are under tension to return to an expanded condition, and this wall tension, plus the water outside trying to get in, create pressure on the door, which prompts it to fly open when released by a touch on the trigger hairs.

These rather bristly trigger hairs project from the door, among the stalked glands. They are centrally hinged, and the slightest touch on one of them by any blundering creature will open a chink around the door so that it snaps open, the expanding walls drawing water in with a rush, and carrying the prey in with it. The door closes again by regaining its former shape.

Then the process of expelling water through the walls of the bladder begins, but sometimes this takes a considerable time. It has been timed in a number of species, and varies from fifteen minutes to two hours. The fact that this is by osmosis means that the water in the bladder must be affected in some way by the plant or its prey to make it hypotonic, so that it passes through the bladder walls to leave the bladder empty.

Although there are short-stalked glandular hairs inside the bladders, there appear to be no true digestive glands, so there is no active digestion; but once the prey is dead, it decomposes. Bacterial action may play a part in this decaying process, but some scientists doubt this because of the presence of benzoic acid. Eventually the products of decomposition are absorbed by the walls of the bladder. In spite of this pattern of decay and absorption, some small organisms manage to live in these traps, among them being algae, infusorians, euglenas, heteronema, phacus, diatoms, and desmids.

According to some collectors' records, *Utricularia* has been found in some strange places. For instance, as well as in relatively deep water, it has been found growing in "vases" (see Chapter 7). Vases are plants in which the leaves form a very close-fitting arrangement at their base, so close, in fact, that the vase thus formed acts as a water reservoir and imprisons small animals that somehow find their way there. Leaves, insects, frogs, and lizards are collected in them, and the water and dissolved chemicals are said by some investigators to pass into the leaves of the plants; but there is only slender evidence of this.

In Figure 2, which shows the action of a bladderwort trap[71], *a* shows a small water animal touching one of the trigger hairs. In *b* the walls have begun to expand, the door has partially opened, and the animal is being drawn into the trap with the inrush of water. In *c* the

2. One of the commoner bladderworts is *Utricularia vulgaris*. An effort is made here to show some of the structure of a trap *(left)*. The traps are also shown on the foliage *(right)*, which has a flowering stem. Also shown is *a* a small water creature touching a trigger bristle, *b* being drawn through the door, *c* well inside the trap, and *d* with the trapdoor closed again.

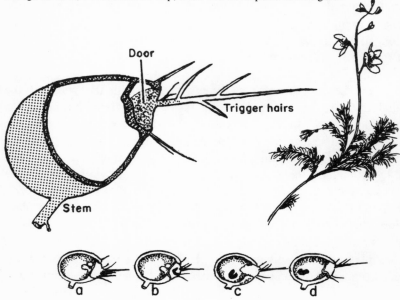

animal is right inside the trap, and the walls are fully expanded, whereas in *d* the door has closed again, and escape is impossible for the prey.

Another genus of plant with traps similar to those of the smaller species of *Utricularia* is *Biovularia,* of which there are two species. Both have the same kind of triggering mechanism as the one just described[71], but the traps themselves are more like those of another slightly different species[70]. These traps have short, tubular entrances, and a tubercle centrally placed on each door carries six trigger hairs or bristles, which bear single-celled mucilage glands. Outer surface hairs also have both mucilage cells and *sessile* ones (attached at the base). There is one great difference between these plants and the utricularias, however: each trap consists of the entire leaf. It is on a stalk, and it is very small, often less than a millimeter in length.

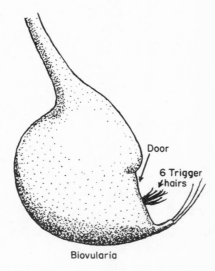

Biovularia

3. The *Bioluvaria* trap is triggered in the same way as that of *U. vulgaris,* but the trap itself is similar to a different species.

Polypompholyx is another of these plants with traps on stalks somewhat like those of *Utricularia,* but its traps are often as much as four millimeters long. There are only two species of *Polypompholyx,* both found in wet sandy soil, and one is illustrated in Figure 4. Another way in which *Polypompholyx* differs from *Utricularia* is in

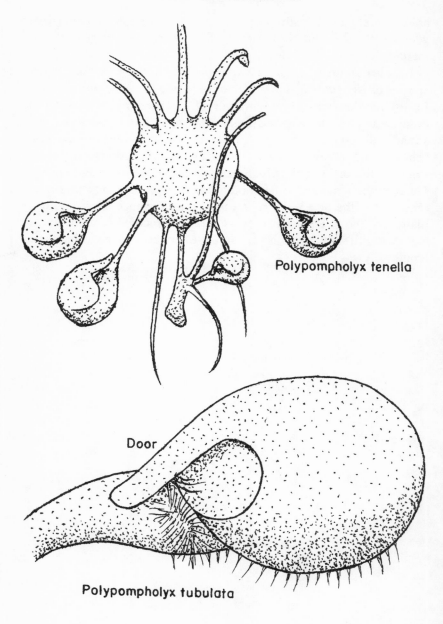

Polypompholyx tenella

Door

Polypompholyx tubulata

4. The traps of *Polypompholyx* are on stalks, and are quite different from those shown in Figures 2 and 3.

having thirty or forty hairs in front of each trapdoor; otherwise the trapping mechanism is very similar.

There is a single species of small freshwater plant[4] which is related to the sundews. *Aldrovanda* is widely distributed throughout the warm temperate and tropical areas of the world, from southern France to Japan, and south to Australia and the African southern tropics. In some respects this plant's traps work rather like those of the Venus's-flytrap on land, because prey is caught between the two halves of a leaf which are concavo-convex in section. This is an extension of a wedge-shaped leaf stalk about six millimeters long and four millimeters wide. The whole plant is ten to fifteen centimeters long and about two centimeters wide, and its flowers are about eight millimeters in diameter, so it is not a large plant.

Some experts think this plant is more highly specialized than any other for capturing prey. At least it compares well with the bladderworts. It is small, and like *Utricularia* it is rootless and lives in swamps, or just below the surface of fresh water. It is usually found among the stems of other plants that seem important to its existence, possibly because they are the breeding sites of much of its prey. This habit of growing among other plants hides it from all except very observant people.

One of the interesting things about it is that it can survive easily without its carnivorous habit. It can, in fact, be grown in the alien environment of a specially prepared solution of mineral salts. However, even though it is difficult to prove that digestion and absorption of small creatures are essential to it, it is certain that the plant does supplement its nutrition in that way. Its degree of carnivorousness may vary with the mineral content of the water in which it is growing, and this may well change from time to time.

Aldrovanda's wide distribution in the freshwater swamps and lagoons of so many countries has made it familiar to many botanists in spite of the fact that it grows among other plants. In Australia it has reminded them of the bottlebrush flower in its formation, but as will be seen in Figure 5, there is really no similarity between the two when their structures are examined and compared closely. Its leaves are like the spokes of a wheel, and these whorls of leaves are set closely along the main stem, like so many wheels on an axle.

The eight leaves of each whorl are pivoted to one side, and all face outward in the same direction. At a touch, every leaf blade can fold together like a spring trap, but a spring trap that is almost too

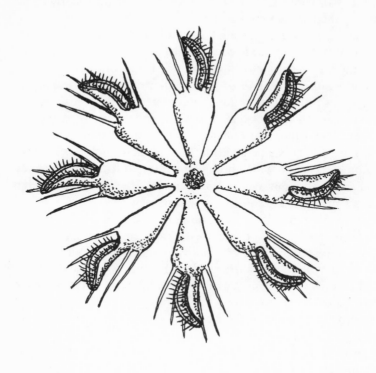

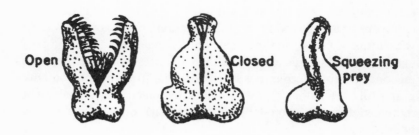

5. An underwater trap that works like the Venus's-flytrap does on land is found in *Aldrovanda*. The mechanism is slightly different from that of the land species, but the principle is the same. The leaves are whorls of eight, and they can all function simultaneously if enough food is available. At the bottom can be seen the various stages of trap closure.

complicated to describe. The blades of the leaves are thicker along the midrib and thinner toward their margins. The midrib is not a hinge, and the thinner parts of the leaf can fold together to meet each other above the midrib by bending.

The outer surfaces of the leaf blades have a greater surface area due to folding, and this facilitates the bending, but how this is stimulated is still not certain, even though many experiments have been carried out using electrical current, chemicals, mechanical and touch stimuli, and organic substances as activators. There are digestive glands on each leaf surface, and near the midrib are about forty long, sensitive hairs that act as triggers when they are touched. The leaf edges have interlocking spines like the Venus's-flytrap[23] (see also Chapter 5).

The shutting of the leaf trap takes a fiftieth of a second, and besides imprisoning the prey, this expels water. Squeezing then follows and continues for several hours. Prey are usually daphnia and water spiders, but sometimes larger creatures, like small fish, are taken in too, and these take some time to consume. As with numerous other carnivorous plants, too much feeding causes deterioration and death of the traps. This plant has a few single white flowers that project up into the air for pollination.

A most interesting underwater trap, of which there are ten known species, is that of *Genlisea*[37]. These plants inhabit swamps in South America, and there is one species in the wet grasslands of western tropical Africa. They live submerged in shallow water with only their flowers projecting above the surface. They have no roots, and their flower structure is almost the same as that of *Utricularia*. All *Genlisea* species are rosette plants with two kinds of leaves: one tubular kind for trapping prey; the other being ordinary foliage. In one common species[37] the trap is only one millimeter long by 0.7 millimeter wide, but other species have traps up to five times this size.

Figure 6 shows how the bulblike trap sits on a tubular neck that makes it look like a Chianti bottle. Inside the bulb there are two ridges, one dorsal and the other ventral, which carry glands presumed to be of the digestive-enzyme-secreting kind. The inner surface of the tube has about forty transverse ridges, each bearing stiff curved bristles projecting downward and inward, and making the tube like an eel trap, or, more accurately, a series of traps. Each section thus formed has two transverse rows of glandular trichomes.

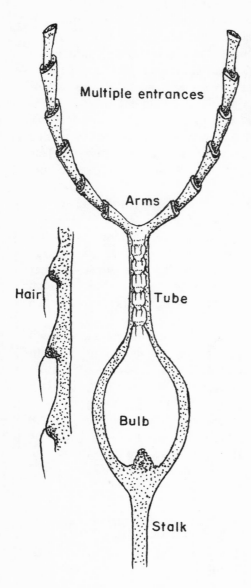

Multiple entrances

Arms

Hair

Tube

Bulb

Stalk

6. *Genlisea* has very unusual underwater traps with many entrances on two spiral arms of each leaf. The prey is eventually passed down into the tube below, and then into the bulb for digestion. On the left, a small section of the tube is enlarged to show downpointing hairs that prevent escape.

At its open end, the tube widens and divides into two arms, which spiral and contain more downpointing hairs. Each of these spirals appears to provide an additional entrance for prey, so that no matter from what direction it may approach, it always has an inviting entrance before it. And these plants have a wide variety of prey: copepods, small water spiders, nematodes, and many other small water creatures. They all reach the bulb eventually, but usually they are well digested before this, probably because there are so many glands everywhere. In spite of its unusual structure, this plant is a very effective trap, and it catches a considerable amount of microscopic life.

3. Pitfall Traps

*M*any plants have their leaves modified into hollow vessels resembling pitchers, and this is how they came to be called "pitcher plants." These pitchers hold a liquid secretion, or a mixture of rainwater and secretion in which insects drown, and which then breaks down the animal tissue into digested products that can be absorbed by the walls of the pitchers.

In many, the rising walls of the pitchers are sloped inward toward their mouths, and this, combined with a slippery substance secreted by glands in the walls, prevents the insects from gaining a foothold that will enable them to escape. One study has shown that certain pitcher plants have walls covered with minute scales that stick to an insect's feet so that they cannot grip the walls, but slide every time the insect attempts to get a foothold.

There are two main groups of these plants: one is found in Southeast Asia, and the other is common in North America. The first group, known as *Nepenthaceae,* was originally discovered in Madagascar in the seventeenth century. Now, about fifty species are known in Ceylon, Southeast Asia, south China, New Guinea, northern Australia (Cape York area), and the Philippines. They are apparently adapted to all climates, for they are found at sea level in

Pitchers

7. The pitchers of the many species of *Nepenthes* range in size from two to eighteen inches in length. They are modifications of the leaves. The plants can be found in many different kinds of climate and location, and altogether there are about fifty species.

the tropics, up on the highest mountains, and sometimes in quite arid regions, where they are short-stemmed with rhizome roots. Those growing in warm, wet areas have creeping rhizomes and often climb to the tops of tall trees.

Although it is obvious that the pitcher of *Nepenthes* is a modification of a leaf due to a continuation of the development of the midrib (see Figure 8), there is still considerable uncertainty as to the way in which this formation differs from other leaf development. The midrib at the leaf tip extends into a tendril, which in turn produces the pitcher. One of the accompanying illustrations (Figure 13) shows ribs on the outside of another species of pitcher. These are rigid, and they end in downpointing teeth within the pitcher after they have turned over the rim. This further impedes the escape of any insect that goes past them.

Around the neck of any pitcher there are nectar-producing glands, and the secretion from these attracts insects over the rim. It also makes that particular zone of the sides of the pitcher slippery. Farther

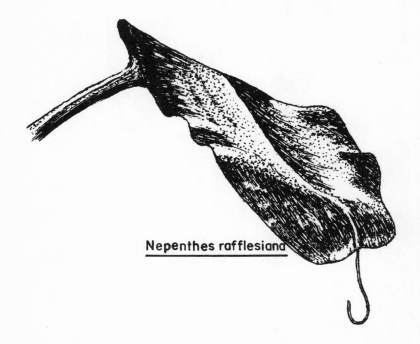

Nepenthes rafflesiana

8. The leaf of a *Nepenthes* showing the tendril growing out from its tip, on the end of which the pitcher will form.

9. Here *(left)* is a *Nepenthes* pitcher in the juvenile stage of its growth, that is, before the lid has opened. The mature pitcher on the same plant *(right)* shows the lid fully opened.

down, the walls become "unwettable" and even more slippery, probably from a waxy secretion. Below this area are enzyme glands whose secretions digest tissues, whereas at the bottom is the liquid, probably rainwater, into which the protein-digesting enzymes flow. The extent of each zone occupied by the different kinds of glands varies in different species.

An interesting feature of the digestive fluid that was recognized by Darwin in his experiments is that apparently it has no power to digest before it has been stimulated by an insect or some form of prey whose touch automatically sets the glands secreting.

Digestion is no doubt assisted in some pitcher species by the action of decay bacteria, which produce simple nitrogen compounds that the plant is able to absorb. Insects first enter the pitcher on foot, attracted by the nectar. They slide downward, unable to fly out because of the narrowness of the pitcher throat, and they eventually fall into the enzyme lake beneath them. One very interesting feature of this liquid is that in spite of its power to digest insects, minute forms of water life and mosquito larvae have been reported as living in it, and in some species an occasional tadpole, perhaps from an egg dropped from above by a bird. This suggests that the enzymes may have some very specific quality that reacts only to certain forms of life, and it casts some doubt on there being any similarity in the digestive actions of these and underwater carnivorous plants.

The Nepenthaceae family of plants includes herbs that grow in bogs or swamps, and *epiphytes,* plants that grow on other plants, but not necessarily as parasites. They climb with tendrils that are extensions of leaves, and which also bear the pitchers. Most pitchers have lids that are sealed in the incompletely developed juvenile stage (see Figure 9), but which open later (Figure 10). These lids will continue, however, to act as umbrellas to prevent too much water from entering the pitchers and diluting the secretions of the various glands too much.

Before the lid opens up, it adheres to the walls of the pitcher by a rim of densely interwoven fibers like a cotton pad, and at that time it appears as though the lid and the main body of the pitcher were quite continuous.

Pitchers come in many sizes and many colors: green with whitish, violet, brown, or crimson spots; creamy whitish-green, and many shades of red. These colors may be the first thing to attract insects to the pitchers, and then, when they scent the nectar, they enter. This suggests that the pitchers are selectively equipped to attract flower-pollinating insects. In addition to different sizes and colors, pitchers will often be found to have quite different shapes even on one plant.

One species[43], *Nepenthes khasiana,* in India grows pitchers seven inches long and three inches in diameter. Another, *Nepenthes maxima*[44], in Borneo has pitchers up to eighteen inches long. It is these that could prove dangerous to nectar-drinking birds. *Nepenthes rajah*[46], also found in Borneo, is said to grow leaves and pitchers which together are up to six feet in length. There appears to be no careful record of the insects the various pitchers attract, but one might assume that the different sizes have evolved to attract different sizes of insects.

There are legends of travelers being saved from death by thirst when they found large pitcher plants holding quantities of water, and

10. A side view of a pitcher showing how the lid will protect the mouth opening from excess rain entering it and overdiluting the digestive juices at the bottom.

11.   The regular multiple blossoms of one species of *Nepenthes*.

Dark amber blossom

Nepenthes rafflesiana

this may be true, but it is also a fact that few such plants grow in terribly dry regions of the world.

These plants are not always in bloom at the time they are found and photographed, so most of those discussed in this book and shown in photographs lack blossoms, which were then drawn separately; and the blossom of one[45], is shown in Figure 11. These blooms are arranged on short, nearly equal individual stalks at regular intervals on an elongated main stalk. They are either male or female, and colored in several shades of red, yellow, and green.

Cephalotaceae is a family of pitcher plants of which there is only one species[14], and it is shown in Figure 12. *Cephalotus* grows from underground rhizomes in the peaty swamps of King George's Sound on the southwest coast of Australia. It is usually found on small elevated tussocks in dense scrub and reeds, and the pitchers are often so brightly colored (usually purple) that they are mistaken for flowers. The real flower is, however, a very inconspicuous small whitish-green affair with no petals, growing on a stalk about two feet high. But although it is so small, it has an attractive sweet perfume.

The pitchers of this plant are up to two inches long, and the pitcher base rests on the ground instead of hanging, as in *Nepenthes*. The immature leaf has a closed lid like *Nepenthes,* and in the adult plant this opens like an awning to protect the pitcher from overfilling with rainwater. In summer it closes in dry and opens in humid conditions.

12. **Cephalotus** is a short ground pitcher plant, the traps of which are rather squat, but highly efficient. They are so tightly bunched that it is difficult to see their real shape, so Figure 13 has been drawn to show one of the pitchers in section.

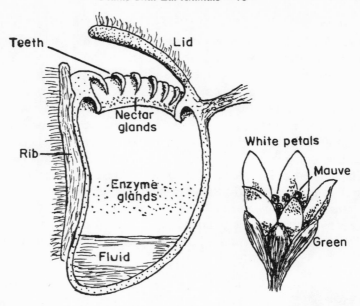

13. Section of a pitcher of *Cephalotus follicularis,* and beside it a blossom from the same plant.

As seen in Figure 13, each rib curves over the pitcher edge and turns downward in a pointed tooth. Nectar glands on a smooth white ledge, or collar, directed downward over the cavity of the pitcher, encourage the entry but stop the exit of insects. Below this are digestive glands. When growing in shade, the pitchers are green, but in the sun they are purple or red. Tiny transparent larvae are reputed to live in the fluid at the bottom of the pitcher and help to consume the insects, just as in *Nepenthes*.

Sarraceniaceae, the name given to a family of North American pitcher plants, was originally derived from the shape of the plant, which reminded observers of a Saracen's helmet. There are about ten species of the genus *Sarracenia,* which can be found widely distributed in the wet swampy areas of the eastern United States and southeastern Canada. *Sarracenia purpurea,* for instance, ranges from Labrador to Florida. There is also one in the central bog lands of western Ireland which, like the other places, is poor in nitrogenous products. Several of the *Sarracenia* are shown in Figures 14 through 17.

Imaginative common names have been given to some of the *Sarracenia* species, including "boots," "huntsman's cup,"

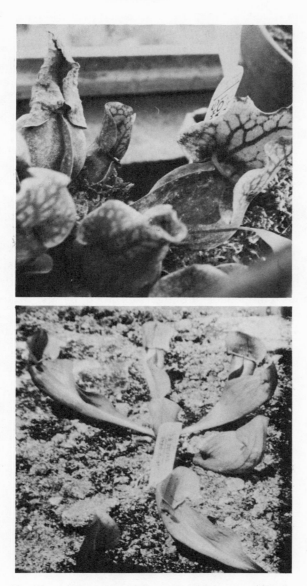

14.   There are ten species of *Sarracenia*, all except one confined to North America. *S. purpurea*, shown here *(top)*, has pitchers that are very colorful with bold veining.

"sidesaddle plant," "soldier's drinking cup," "forefathers' cup," "trumpet leaf," "Indian cap," "watches," and "trumpets," according to the impressions their shapes have made on different observers.

Although the pitchers of *Sarracenia* are rather similar in design, structure, and function to those of the *Nepenthes,* the plants are not closely related. They belong to widely separated families. The trap of a *Sarracenia* also has an upper region with many nectar-secreting glands. The nectar increases in density as the horn of the pitcher narrows, and this lures insects farther in. Then they find that their feet cannot grip the down-pointing needlelike hairs, which, together with the narrow tube, prevent their escape.

It has been suggested that these plants produce something to make insects dopey, because they do seem to become slow and sleepy before they finally drop to the bottom of the pitchers.

Once again, proteolytic (protein-digesting) enzymes are secreted into the rainwater that collects at the bottom of the pitchers, and this mixture brings about decomposition of the prey. The resulting nitrogenous compounds are then absorbed by the walls of the pitchers.

There always seem to be astonishing exceptions to everything that

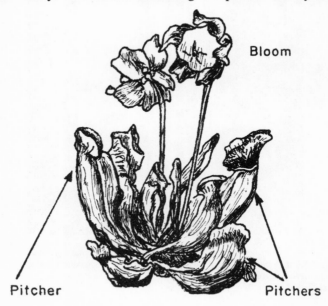

Bloom

Pitcher

Pitchers

15. The blossoms of some species of *Sarracenia* are unexpectedly large and colorful.

16. Some of the *Sarracenia* pitchers are very large, up to two feet long and three inches wide.

happens in nature, and there is an example of this found in these pitchers. An unusual species of mosquito lays its eggs in the pitcher water of *Sarracenia,* and this insect has adapted itself successfully to living out its larval life in that fluid. When larvae finally hatch into adults, they climb out of the pitcher and fly away with no trouble at all. Their feet do not slip on the pitcher walls, as those of other insects do, and so far no one has been able to discover why this is so. It may be because their feet have not previously trodden in the nectar, but

Dark crimson blossom

17. Another *Sarracenia* pitcher with its exotic blossom.

Sarracenia mitchelliana

even this would not explain why the larvae can live in the pitcher water when other insects will be consumed. The mosquito larvae live on remnants of the pitcher's prey until they metamorphose.

These pitchers are brightly colored, ranging from reddish-purple to green with purple spots, greenish-yellow, red-white-green, and other combinations; and they have lids that function exactly the same way as was described for the various *Nepenthes*. The plants have single inverted blossoms on long stems. They are usually purple, and are bisexual.

The photograph of *Sarracenia purpurea,* in the upper part of Figure 14, has pitchers up to six inches long, green with red streaks, and with purplish-green flowers. *Sarracenia flava,* in Figure 16, has pitchers up to two feet long and three inches wide, and its yellow flowers can be five inches in diameter. *Sarracenia psittacina* pitchers are red mottled with white fenestrations.

In California there is a swamp species[22], which is a little difficult to describe, but which is shown in Figures 18, 19, and 20. The pitchers have inverted entrances and hoods that resemble fish tails. These hoods, which have nectar glands on their undersides, have been compared to snakes' tongues, and with the suggestive shape of

18. *Darlingtonia californica* has a very unusual pitcher. It is turned downward, and grows in clumps.

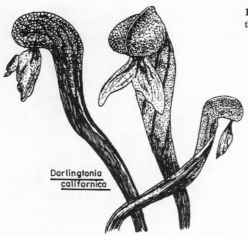

Darlingtonia
californica

19. An effort to make the details of pitchers like those in Figure 18 a little clearer.

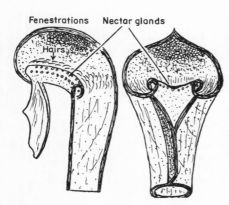

Fenestrations Nectar glands

Hairs

20. A vertical section through a pitcher of *D. californica,* and at the right a frontal section through the head.

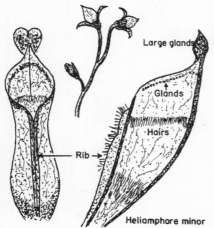

Large glands

Glands

Hairs

Rib →

Heliamphore minor

21. *Heliamphore* is very much like some *Sarracenia* species. It is boldly marked with red veins, and has small white flowers. The five known species vary greatly in size.

the head of the pitcher, this species is popularly called "cobra plant." The green tubular pitchers are scaled with white at the top, and have reddish veins. The flowers are pale green with reddish-yellow petals.

There are spiders in some parts of the world that build their webs inside insect-trapping pitchers. In this way they catch the insects that fall into the pitchers and so rob the plants of at least some of their prey, but it is not clear how they manage to fasten their webs to the pitcher walls, or how they manage to survive themselves.

Heliamphora belongs to the family of Sarraceniaceae. There are only five species, all found in British Guiana and Venezuela, but it is evidently a hardy group, because one species[40] has been found growing at six thousand feet above sea level. It grows in the wettest places, in regions of vast rainfall, and perhaps for this reason—and because it likes open places where the grass is short—it is one carnivorous plant that is not easy to raise at home.

These pitchers have no actually slippery zone, but the area is smooth. They are red-veined, and the plant's white flowers rise on red-tinted stems. The entire leaf, including the pitcher, may reach a length of twelve inches, and one species[41] may grow to sixteen to twenty inches.

The pitchers of *Heliamphora* grow in clumps and have the usual nectar glands and down-pointing hairs. In this respect they are very much like species of *Sarracenia*; but here the similarity ends. Heliamphora has no digestive proteolytic glands, and prey must therefore be broken down by bacterial decomposition before it can be absorbed, which is a slower process than when enzymes are present. This sets the genus apart from other pitcher plants already discussed.

There is an area of swamps and bogs in Texas known as The Big Thicket. It lies between Segno and Beaumont, and four of North America's six genera of carnivorous plants can be found there by anyone who cares to search patiently. The sundews *(Drosera)* and some of the pitcher plants *(Sarracenia)* are there, and the latter are particularly good. The nectar secretion on the underside of the hoods is clearly visible, and when the sun shines through the semitransparent pitcher walls, insects can actually be seen inside them. And insects are not all that these plants consume. Occasionally a small skeleton may be found in a pitcher, when some luckless lizard has become trapped, and its entire body tissues except the bones have been completely absorbed by the plant's digestive enzymes.

4. Flypaper Traps

Sticky leaves that hold insects until they are digested and absorbed are found mostly in a family called Droseraceae. Of these there are probably a hundred species, certainly ninety species at least of the *Drosera* genus, all dicotyledonous: the seedling forms two primary leaves; there is one known species of *Drosophyllum*, which is monocotyledonous: the seedling forms a single primary leaf.

Drosera species range through California, Europe, and the North Temperate Zones, as well as South Africa, Australia, and New Zealand. The one species of *Drosophyllum* is confined to one specific dry area of the world, southern Portugal and Spain, and Morocco. Many of the *Drosera* species can be raised successfully at home, and how this can be done is described in Chapter 8.

Sundews. *Drosos* means "dew," and these plants are named sundews because of the way the droplets of their secretions sparkle in the sun. This secretion is the most powerful adhesive in nature, and seldom can any insect free itself from the plant. They are herbs which, except for *Drosophyllum*, ordinarily grow in bogs where the soil is acid and deficient in nitrogenous compounds. Any soil that never dries out usually has this deficiency, and this is undoubtedly why these plants need animal protein.

Although most of the sundews are small and relatively inconspicuous, they are very effective insect traps, the insects probably being attracted by their dewlike sparkle as much as by their alluring secretion. One species[34] in Australia may have up to a dozen insects on any one leaf at a time. Most will be flies, but butterflies and even dragonflies have been known to become trapped on sundews.

The leaves of most of the *Drosera* species usually grow in rosette

22. Sundews sparkle brightly, the drops of secretion on their adhesive stalked glands giving the illusion of dew. Captive insects can be seen on three of the leaves. Close-up shows an enlarged leaf *(right)*.

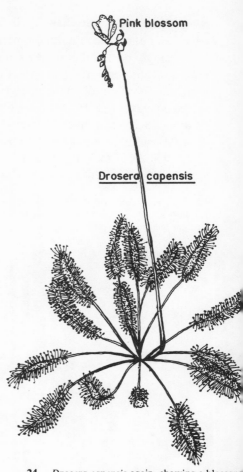

Pink blossom

Drosera capensis

23. Two different kinds of sundew, *Drosera capensis (above)* and *Drosera rotundifolia (below)*. Captive insects are visible on the upper plant.

24. *Drosera capensis* again, showing a blossom

form, and Figures 22 through 26 show this, but they vary greatly in shape between species. They may be long and narrow or short and oval. Some of them are even forked. One species[30], shown in Figure 27, is quite unlike any of these, however; the circular leaves are scattered along irregular stems.

The leaves have stalked or tentacle glands on their upper surfaces and margins, and these are large, often reddish in color, producing both an attraction odor and the globules of mucilage to which the insects become stuck. The centrally placed stalked glands are short, whereas the outer ones are much longer and very flexible, and one leaf may have from 130 to 250 of these glands.

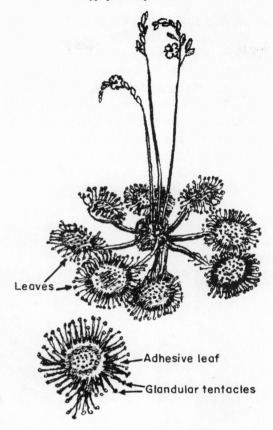

Leaves

Adhesive leaf

Glandular tentacles

25. *Drosera rotundifolia* with its blossom.

The tentacles are so sensitive that the very slightest touch activates them, and it almost seems as though they have a nervous network that communicates this touch to the other tentacles around them to make them bend over toward the prey and smother it. Raindrops falling on them do not stimulate them, however. On both surfaces of the leaves there are sessile glands which secrete the proteolytic enzymes that digest and absorb the captured prey.

The glands and the tips of the stalks are also sensitive, and it has been suggested that it is the leaves' fluid movement which transmits stimulations from one stalk to another. But in some species, if the stimulus occurs on one edge of a leaf, only the stalks on that side are affected. If most of the outer stalks become inflected (turned in), the leaf cups itself to form what might be considered a panlike stomach into which the digestive enzymes are poured. If nonorganic material is dropped onto a leaf, this will cause less contraction and of a shorter duration than a stimulus caused by organic material.

An insect alighting on a leaf is held by the sticky secretion on the tentacles, its presence stimulating a continuing secretion of this substance, as well as causing the surrounding tentacles to move toward it until it is enveloped in a gluelike mass. This is very clearly seen in one species[26], *Drosera intermedia*. This movement of the tentacles increases the firmness with which the prey is held and

26. *Drosera aliciae* grows in tight rosettes. Several small insects are visible on the leaves.

27. The "erect" sundew, *Drosera peltata,* is quite different from all the others shown. The arrow shows a leaf with its hair glands and "dew."

makes the plant something of an "active" trapper. At times, two leaves may combine forces to hold a large insect in *Drosera rotundifolia*.[34] A single leaf may also curl itself right around an insect. In sundews with long leaves, only the tentacles close over prey, but those with round or oval leaves curl an entire leaf around the prey like the Venus's-flytrap, unfolding later to reject the insect's useless husk. It seems a leaf succumbs rapidly to the bite of an ant.

Drosera rotundifolia often grows prolifically where there is food all ready for taking. It thrives on floating sphagnum, and mosquitoes lay their eggs in the water beneath and around it. The mosquito larvae come to the surface of the water to hatch into flying adults, and the first thing they do after hatching is to climb up a plant stem. Nine times out of ten the plant will be a sundew, and the odor of the nectar attracts them to the leaves before they fly. Before the insects have been adults for more than a few minutes they are struggling to their deaths on the sticky tentacles of sundew leaves.

Digestion is complete in several hours, and then the stalked adhesive glands relax and expand, resuming their original positions. The undigested wings and other debris are carried away by the wind before secretion begins again and the leaf becomes ready for fresh prey.

Sundews have been given many popular names that describe them well enough for the nonscientist. "Pygmy sundews," for instance, describes some very miniature species[28, 33], which have dense rosettes of small leaves with long slender stalks, fibrous roots, and flowers either white or red. The "western sundew"[29] is a pygmy with a single white flower. The "pearly sundew"[33] is another pygmy with six to eight white flowers. The "shining sundew"[31] and the "wheel sundew"[27] are all pygmy plants.

"Rosetted sundews" describes larger plants with flat rosettes of leaves, and leaf stalks that are almost indistinguishable from the leaf blades. These plants have large bulbous roots. "Fan-leaved sundews" grow erect, with fan-shaped leaves along their stems. "Rainbow sundews" are also erect, and may even climb. The leaves are cupped and situated most frequently in groups of three on slender stalks. "Narrow-leaved sundews" have long, narrow leaves, as the name implies.

The "round-leaved" or "common sundew"[34] is shown in Figures 23 and 25, which well illustrate it. Of the many species of sundew in the world, at least fifty-six are found in Australia, 80 percent of these in the southwest of the country, and about 15 percent in the east. Many parts of Australia suffer from periodic droughts, and in Western Australia sundews appear to die in hot, dry summers, but some survive with tubers, and others because they are protected by the dead rosettes of their leaves.

Rain brings out new leaves, and probably the necessary insects too. Sometimes small rosettes grow on the leaves, and these drop off to take root in the ground close by.

Sundews have an immune insect living commensally (in partnership) on their stems and leaves, the "assassin bug," which matches the plant leaves in color and markings. The plant's secretions do not affect it, and it sucks out the body contents of the plant's prey through its powerful proboscis, which first injects a venom that kills the insect in seconds through a neurotoxic action. To what extent this bug is beneficial to the sundew is not obvious, unless its very presence attracts other insects. But though it seems to have everything in its favor, there is always the hazard in some places that the leaves of the plant die in spells of dry weather.

Drosophyllum lusitanicum is very different from all the other sticky flytraps. One expects to find all carnivorous plants living in swampy or damp locations, or in bogs where the soil is poor in nitrogenous products, so it comes as a surprise to find one growing in dry rock crevices or in very dry soil on hillsides. This plant does just that in the near-desert areas of the coastal regions of southern Portugal and Spain, and in Morocco; and one can assume that in these locations too it would be difficult for plants to obtain nitrogen from the soil. The plant is sometimes found sitting on a dense mat of its own dead leaves, or perhaps on top of a clump of heather, or even in the undergrowth of cork woods.

As shown in Figure 28, this unusual plant has long, thin leaves, which will prevent dehydration to the greatest possible extent. This is important because the temperature in these areas can easily reach around 125°F in the shade, and very much more in the sun. In fact, southern Spain, especially the southwest, is known as "the frying pan of Europe." The leaves of *Drosophyllum* are covered with purplish sticky hairs and stalked glands, which, like *Drosera,* give off a sweet aroma that attracts insects. But they are not mobile like the stalks of *Drosera,* and their secretion is different; it is more acid, and it is secreted continually.

The several bright-yellow flowers on a single stem about a foot tall are up to an inch and a half in diameter. The sessile glands, those which secrete the digestive medium, are like those on other sundews; they are emergences, which means they grow from tissue below the epidermis; and they only secrete when prey is actually caught, so they function differently from the stalked glands. Their secretion seems to be triggered into action by contact with prey by the stalked glands, which is similar to the action of a number of other carnivorous plant leaves.

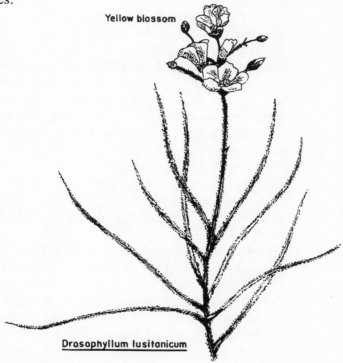

Yellow blossom

Drosophyllum lusitanicum

28. Quite unlike any other kind of sundew, *Drosophyllum lusitanicum* grows in dry desert or rocky areas, but is an excellent flycatcher, hung by local inhabitants in their houses for this purpose.

Drosophyllum catches unusually large numbers of gnats and flies. Often twenty can be seen on a single leaf, and it is said that people in the areas where these plants grow hang them in their cottages as living flypapers. Digestion is relatively rapid, twenty-four hours being all that is necessary for its completion, and this applies to any insect up to the size of a mosquito. Because there is formic acid present in their secretion, the digestive glands of this plant discourage the presence of decay bacteria, but this is so in a number of carnivorous plants.

There is a larger-than-usual flypaper plant in the western part of the Cape of Good Hope[56], which is rather like *Drosophyllum*[35], but not in its fine detail. It is included with the rainbow plants, and it grows to a height of several feet under favorable conditions. Its leaves are covered with tentacles and glands of many sizes, which, although they are not mobile, capture many insects.

Rainbow Plants.   Except for the large plant just mentioned, rainbow plants belong to the family Byblidaceae, and there are only two species of this kind of flypaper trap[12, 13], both found only in Western Australia. They are very much like *Drosophyllum,* but perhaps less shrubby, and they grow in different kinds of localities, usually sandy swampy areas where one is also likely to find *Polypompholyx spp.*[55] and some species of *Utricularia. Byblis gigantea,* which grows in the drier areas of the swamps, may reach a height of twenty inches. The yellow-green leaves are triangular in section and are covered with stalked mucilage glands as well as nectar glands on all sides. The usual digestive glands are in rows closer to the leaf surfaces. They are not sensitive like the leaves of many other carnivorous plants. Their flowers are violet or pink.

These rainbow plants have a capsid, or wingless leaf bug, living commensally with them, just as the clown or damselfish[5] does with the sea anemone; and like the assassin bug does on *Drosera*. It is quite immune to the secretions of the glands, and it shares the prey of the leaves, extracting their juices; but what it does for the plant in return, apart from luring other insects that will become prey, is no clearer than with the commensal partners of other carnivorous plants.

Butterworts.   Finally we have an interesting carnivorous plant with roots and a basal rosette of leaves that has been put to good use by man in some areas of the world. This is the genus *Pinguicula*. There are about thirty species, all in temperate regions of the Northern Hemisphere. The name comes from the word *pinguis,* which means "fat." These plants are found in damp places on moors and mountains where the soil is acid, and also on boggy ground.

Also known as "bog violets" because their flowers vary from white through mauve to purple, the butterworts derive their name

from the use of their leaves for curdling milk, a property that appears
to be present in the proteolytic enzymes associated with the plant's
carnivorous activities. North of the Arctic Circle, nomadic Laplan-
ders have used it extensively for their reindeer milk, and may still do
so, pouring the milk while it is still warm through sieves lined with
butterwort leaves. After a time the milk solidifies, and it can then be
used as butter. The particular species of butterwort used may be

29. These large rosettes of *Pinguicula vulgaris* attract and absorb many
insects, some of which can be seen adhering to the leaves in these photographs.

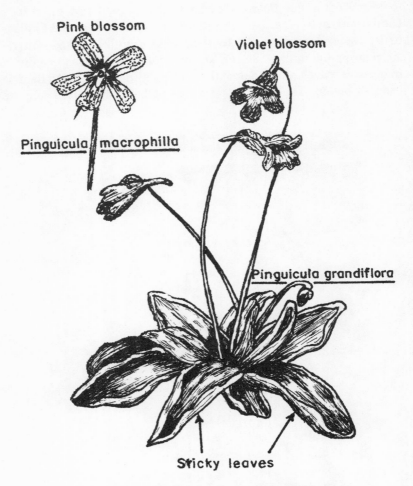

30. *Pinguicula* blossoms are simple but colorful.

Pinguicula alpina, which grows in the region, and which has attractive white flowers with yellow throats.

A common European species[54], which forms flattened rosettes of rather sickly-looking leaves with a yellowish-green fleshy appearance, somewhat oblong in shape, and with upturned margins, is shown in Figure 29. The leaf surfaces have a dense covering of sessile glandular hairs, which are in stalked forms on the upper surfaces. These produce the usual sticky secretion on which small insects get trapped, and the presence of an insect stimulates the production of further secretion, accompanied by a slight inward curling of the leaf margins.

This curling takes from one to two hours to be completed, and lasts for perhaps twenty-four hours. There are also sessile glands slightly below the leaf surface, and their secretion contains the usual proteolytic digestive enzymes, which in this case may be assisted by bacterial action, although no one is quite sure of this. The glands are apparently only stimulated by objects having soluble material of a special nature. All the glands on the upper surfaces of the leaves of this species are reputed to be connected through fine vessels to the vascular system of the plant, and this, most probably, is the source of the material from which the secretions are manufactured.

Species of butterwort on the American continent are very colorful. One[53], which is found in Mexico, has flowers ranging from purple to carmine, and in southeastern North America, another[52] has orange-yellow blooms close to an inch and a half in diameter.

Many other plants have sticky glandular surface hairs, but not all of them can digest the animal tissue that becomes trapped by them. It is thought that some saxifrages may be able to do this, however, and some others seem able to absorb chemicals dissolved in rainwater, but this does not prove them capable of absorbing organic material like animal tissue.

5. Snap Traps

We first mentioned snap traps in Chapter 2. These were the underwater *Aldrovanda*, in which two halves of a leaf suddenly close on prey when it accidentally touches a trigger mechanism. But the best-known snap-trap plant is found on land, and the reason why more people are familiar with it than with any other carnivorous plant may be because it has been given considerable exposure in print for almost a century, in spite of the fact that it is very limted in the territory it inhabits. This plant, like *Aldrovanda*, belongs to the family of Droseraceae.

Venus's-Flytrap. The genus of this plant, *Dionaea*, has only one species[23], which, in spite of its restricted range, has become a very popular plant for growing at home, for it has endless fascination for young people looking for the unique in botanical study. (See Chapter 8.)

Venus's-flytrap is a veritable mechanical marvel, and for all its small size it is probably the most effective and spectacular of all the prey-trapping plants found on land. It is related to the sundews, and is found only in swampy habitats in North and South Carolina, so it is probably also the most restricted species of all those so far discovered.

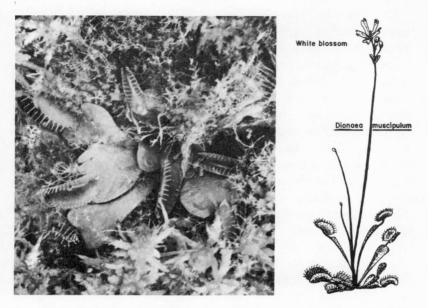

White blossom

Dionaea muscipulum

31. The Venus's-flytrap, most famous of all carnivorous plants, is not large, but it is highly lethal to insects of quite considerable size. *Dionaea muscipula* is shown here with traps in various stages of reopening.

32. The simple blossom of *Dionaea muscipula,* also called *D. muscipulum* by some authorities.

Figures 31 and 32 show the form of this plant, and Figure 33 shows its trapping action. The leaves form a low rosette, growing more or less horizontally and up to six inches across, and they are surmounted by a tall cluster of up to fourteen white flowers. Each leaf stalk is an expanded leaflike structure, and the end of it, which is the true leaf, forms into two parts at the midrib, where they make an angle of 40 to 50 degrees to each other, folding toward each other to form an imprisoning chamber. The chamber is the flytrap.

There are also spines projecting from the edges of the two leaf parts, which further ensure the imprisonment of the prey once it is caught. On each side of the midrib there are three trigger hairs, and when an insect alights on a leaf it can hardly avoid touching these. As soon as it does, the two halves of the leaf quickly fold up, the midrib functioning as a hinge. When rain touches these hairs or any part of a leaf, there is no reaction. The plant is able to differentiate between the living touch of an insect and the impact of a raindrop.

The trigger hairs are also hinged, and they operate only when bent beyond a certain point. One hair must be touched twice or two separate hairs touched at an interval of between two and twenty seconds for the trap to snap closed.

As the two halves of a leaf press the insect between them (see Figure 33), the spines on their edges interlock and prevent it from slipping out or being squeezed out. Tiny insects do escape between the spines, but these would probably be too small to be of much use to the plant anyway. Once an insect is imprisoned, the plant's proteolytic enzymes are released, and it is eventually absorbed into the tissues of the plant. The closure of the two leaf halves can be said to form a temporary stomach.

When digestion and absorption are complete, the halves of the leaf open again and allow the indigestible remains of the insect to be blown away. Then the whole efficient process is ready to start again. The reopening of the lobes seems to be a response to expansion, perhaps by growth of their inner surfaces, and this takes up to ten days to complete unless there has been no effective capture, as when a tiny insect escapes between the spines of the leaf edges.

Insects will not alight readily or frequently on a leaf unless they are attracted, and so, as in other carnivorous plants, the leaves have small

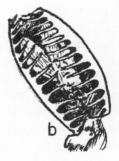

33. In *a* a fly touches down on a *Dionaea* trap, showing *b* the resulting imprisonment, and *c* the ultimate complete closure for digestion of the prey.

glands secreting sweet nectar, but these are located only along the outer margins of the two halves of each leaf. The glands that secrete the digestive enzymes can be seen on the inner surfaces of the leaf halves as pinkish or reddish areas.

These digestive glands are sessile, and they are larger and more prominent than the nectar glands. They do not secrete when nonanimal material causes the trap to close, and they remain dry and inactive until real prey is captured. Then, after releasing the digestive enzyme, they reverse the process and absorb the products of digestion. Between the enzyme glands and the nectar glands at the leaf margins there appears to be an entirely gland-free zone.

The triggering mechanism that responds to the touch of an insect, the small hairs on each side of the midrib, are activated by two or more stimuli, either of a single hair, or of different hairs. An interesting thing about this is that if the stimuli are weak, perhaps from a very small insect, they add up over a period of minutes until the leaf closes. But in this case it closes more slowly, unlike the more or less instantaneous closure that takes place when a vigorous insect enters the trap. It has been observed also that the stimulus brings a quicker response as the temperature rises.

Though adhesive leaves might theoretically catch almost any kind of insect, there are definite limitations to the capabilities of the Venus's-flytrap. The size and construction of the trap limits the size of its prey, which is usually no more than a quarter of an inch long. Observers have testified to captures of insects over an inch long, but this must be very difficult for the trap to absorb, and, as with some other plants of this nature, too much food proves indigestible, and eventually fatal. In fact, traps sometimes consume only one or two prey before they become useless, and they may die while trying to digest a third or fourth.

An actual response is possible about once a day, but responses do not always mean successful captures. Even without the deteriorating effects of capture and digestion, the sensitivity of a leaf decreases with age. As with some other digesting plants, the formic acid in the enzymes inhibits the presence of decay bacteria at the same time.

After capture, the trap remains closed for a few days, and the power to close again is not recovered for several days more. Just how closure is brought about by touch is not clear. The stimulus seems to travel through cells quite independently of the vascular system, which was once suggested as the natural pathway. The recording of the electrical wave pattern through electrodes on both upper and lower leaf surfaces shows a disturbance like that seen in the action of animal muscle.

Almost as soon as the Venus's-flytrap has digested insects, the entire plant begins to grow rapidly. New leaves appear and push their way up between the existing leaves, growing higher than these, as though trying to jostle for the best insect-catching position.

The action of animal muscle is in response to nervous impulses, so to find a similar action in a plant, no matter how specialized, is rather startling, but perhaps no more so than the rapid movements akin to reflexes which we see in the various traps used by carnivorous plants. The only things of which we can be certain are that there is rapid movement, and that electrical energy is involved. Just how these come about is still a mystery.

6. Predaceous Fungi

$\mathcal{M}$any people will get some kind of fungus infection at some time in their lives, and most are familiar with the so-called athlete's foot and other infections that can invade the warm, damp parts of the body, especially in summer or hot climates. But almost all would be incredulous that there are fungi that actually trap and consume live animals—small animals, it is true, but active ones nevertheless.

Their prey is often only amoebae, but they can also be as large as nematode worms. These fungi probably confine themselves to microscopic creatures found in soil, and they are quite important in retaining a balance of this kind of life. Their methods of trapping their prey are in many respects similar to those used by the plants described in earlier chapters. Some are adhesive, some have gripping traps, and still others have an adhesive in the trap.

The method of achieving digestion is, however, very different from that seen in the true carnivorous plants. Whereas these plants secrete digestive enzymes that reduce the prey while it is being held captive, or they accept the services of decay bacteria for this, the fungi send out processes into the tissues of the prey and absorb its contents through these.

Such processes are known as *haustoria,* or, in unscientific terms, "suckers." They are also called *hyphae,* and both terms will be seen in the illustrations. Sending out haustoria is not confined to animal prey; it takes place in plants as well, and it is possible that these processes penetrate in response to the moisture within the prey, which is needed by the parasitic organisms.

Fungi can be either parasites (living on other live organisms) or saprophytes (living on dead or decaying organic matter), although they are not necessarily always carnivorous. They exist most prolifically when the soil, the atmosphere, or whatever matter they grow on is damp. Since they contain no chlorophyll they are unable to manufacture their food in the way that green plant foliage can, so they must absorb other organic material, and in certain species this is obtained by capturing living victims. But nutrients diffusing around the growing fungus are also important to its propagation. Without such fungi, which consume microscopic animals and dead animal tissues, the world would soon be choked with dead material.

The structure of a fungus is basically a branching system of hollow tubes secreted by the protoplasm as it grows. These are about three to ten microns in diameter (a micron is a thousandth of a millimeter or a twenty-five thousandth of an inch). One can call the fungus a vegetable structure without vascular tissue, in which there is no differentiation into stem and leaves, and from which true roots do not form.

Fungi reproduce by giving off spores. These differ from seeds of higher plants in that they do not contain an embryo. The propagating spore with which the growth cycle starts sends out one or more tubes, the first being called the first hypha of the new mycelium, as the whole system of hyphae forming the plant body is called. The speed with which these extend themselves is naturally related to the availability of nourishment from the hyphae that are adjacent to food. Protoplasm is synthesized from the nutrients, and new wall material forms around this.

There are two forms of these fungi; one has *septa*, or fine walls, between the hyphae, whereas the other is continuous; it is just a long tubular mycelium (see Figures 34 and 35). When there are no septa the fungus is known as *coenocytic*. The propagating spores occur all over the mycelium instead of being aggregated together in a compact fruit body, as in so many higher plants.

Although what is shown in Figure 34 appears large, predaceous fungi, as already explained, really consist of mere threads just a few microns in diameter, but some of them have amazing structures with which they prey upon the microscopic animals of the soil. In Figure 34, *a* shows amoebae stuck to adhesive processes produced by a fungus. In *b* haustorial processes are growing into the prey to absorb its contents. Then, in Figure 35, *a* and *b* show the operation of a snarelike trap consisting of three cells that suddenly enlarge and grip tightly any nematode worm that blunders into or attempts to press through the ring. In the same figure, *c* and *d* show a species of fungus that forms rings with an adhesive coating on the inside. Nematodes become firmly stuck in this ring. Illustration *e* shows a later stage, when the digesting hyphae of the fungus have grown right into the body of the worm. Another kind of snare is shown in *f,* in which the loops are distributed right along the mycelium.

That such simple organisms should have such complex habits is perhaps unexpected, but man should feel grateful for the part fungi play in retaining a balance in nature. They keep in check many pests that otherwise might seriously interfere with his food supply.

34.   A predaceous fungus, which captures amoebae by adhesion, and then invades the cell's body with digesting branches.

35.   Some fungi trap worms with constricting loops, some of which are also adhesive. Once the worms are well held, the fungus sends digestive branches into their bodies until they are completely consumed.

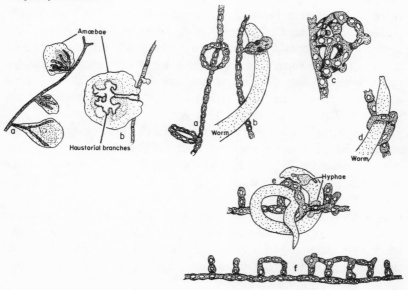

Adhesive Fungi.   One species[1] that captures eelworms has many small loops forming networks, as in Figure 35 *c*, that tend to adopt right angles to each other. These loops are very sticky with a viscid secretion, and any eelworm touching one of them as it presses blindly through the soil is immediately held like a fly on a flypaper.

Branches grow out from the ensnaring loops into the worm's body, the tips swelling to form globular structures called *injection bulbs* up to half the diameter of the worm. From the surface of each of these bulbs small projections grow out into *fungal threads* called *trophic hyphae,* which quickly increase in length until they fill the worm's entire body and absorb its contents, passing this back to the mycelium of the fungus. Something like this process is shown in Figure 35 *e*. It is complete in about twenty-four hours, and then the trophic hyphae withdraw from the carcass of the worm into the body of the fungus network. About twenty of the fifty known species of fungi that capture eelworms do it in this way, but species vary in the degree of their predaceousness[6, 64].

In America and Britain there is another fungus of this kind[19], which is, however, much less elaborate. It has short, sticky branches consisting of one to three cells projecting from the threads of the mycelium. Sometimes these extend into loops, but in a simpler form than is shown in the diagram in Figure 35 *f*. Another species[20], has tiny, almost spherical, sticky knobs close together, very numerous, and each attached to the threads of the mycelium by a very short stalk.

The network kind of trap seems to be more efficient than those consisting merely of sticky knobs, but it is possible that the adhesive secretion of the former contains some toxic substance because the most vigorous worms die long before the fungal threads and hyphae have penetrated into their bodies. There is no certainty of the presence of such a substance, however.

One species[63] is adhesive all over its mycelium, and so does not need knobs or loops in order to be efficient. The hyphae making up its mycelium are not divided into septal cells, but are continuous tubular structures like that shown in Figure 34 *a*.

Ring Traps.   Fungi with *passive rings,* into which worms wedge themselves but which have no adhesive, are called *constricting rings;* see Figures 35 *a* and *b*. Penetrating hyphae consume the contents of the worm in exactly the same way as has been described for the adhesive species. Even if the struggles of a worm break a ring from its stalk, the result is the same. Hyphae still grow into the worm from the ring, and eventually consume it. In fact, the worm's body acts as an incubator for the growth of new rings. Some fungi with passive rings carry a second kind of trap in the form of adhesive knobs on stalks.

In the constricting-ring kind of trap shown in Figure 35 *d*, the cells swell as soon as a worm enters the ring, and this produces a firm grip on it, the entire process taking no more than a tenth of a second, which is almost as fast as the action of snap traps in some carnivorous plants. The merest touch on a cell causes it to swell instantly. Whether this is by some form of osmosis is not clear, but the speed suggests some other way. Without a nervous system, such rapid action is a mystery. The speed of closure of a ring seems closely related to its age, because it slows to a halt as age increases.

When there are no worms in the vicinity, ring traps are not formed, but as soon as any trace of fluid in which eelworms have been reaches the fungi, the loops form rapidly and are ready for trapping in twenty-four hours. Thus there is obviously a response to something secreted by or something within the worms, probably a substance that is very unstable. Moreover, none of these fungi appear to *need* worms. They can get adequate nutrition from other organic substances in the soil, but seem to prefer worms if they are available.

Penetration by Spores.   It seems that the only way predatory fungi can be classified is according to the size and form of their spores and the way these develop. One form, which is rather unique, belongs to certain species called *endozoic* because they enter worms by way of adhesive spores called *conidia*. These are unicellular asexual reproductive bodies, and when a worm brushes one of them, it sticks to the worm's body and more or less hooks itself into it. It then begins to penetrate and form a mycelium with branching hyphae inside the worm's body, which it proceeds to consume.

When a worm dies through having its body contents absorbed by a fungus of this kind [8, 36, 39], the latter projects spore-bearing elements into the air that spread to form new predators as they are picked up by other worms when these brush against the tiny barbs on the spores, and which then penetrate the worms' body surfaces. Not all of these spores adhere to an animal in the way just described. Some are taken in by worms with their food. This happens with two species[15, 16].

Bait Traps.   Some fungi[62] grow short buds on their mycelia, and these must have some attractive quality, because prey try to take them into their mouths. As soon as this happens, the buds enlarge so that the prey cannot let go again. Like the enlargement of cells in some ring traps, this is fairly instantaneous. The buds then project haustorial processes into the animal and consume it as in the other forms already described.

There are so many species of predatory fungi that attack eelworms that it is surprising that any survive, but eelworms and other nematodes produce vast numbers of offspring to maintain their populations. One species of fungus[17] invades the bodies of caterpillars, and there may be other predators of this kind that have not yet been identified.

There are other forms of predatory fungi that are adhesive,[1, 21], and which capture amoebae; and still others,[49, 50], which capture shelled *rhizopods* (protozoan animalcules).

7. Vase Traps

$\mathcal{A}$ny plant in which the leaves are arranged in tight rosettes at the base is likely to catch water and have insects drown in it, but not all of them are capable of using the food these insects might provide. Indeed they will not need it if their requirements are sufficiently met through their roots. There are, of course, many plants with cup-shaped flowers that form deep receptacles for rainwater to collect in, such as the orchid shown in Figure 1. Insects will also be trapped in this, but there is no evidence that these flowers have any digestive action to make use of them.

Figure 36 is typical of the kind of plant that forms the best traps for rainwater and animals. The leaves fit together so closely at the base that they become a vase or pitcher that is as watertight as if it were one continuous unit. Several genera of these collect fallen leaves, insects, frogs, lizards, etc., and their leaves can absorb water and dissolved salts, but to what extent this nourishes the plant, if at all, is not clear. What is clear, however, is that the debris and creatures collected form a ready source of food if the plant can use it.

An interesting feature of these plants is that species of *Utricularia* have been discovered growing in their vases (see Chapter 2). This means that insects and small water creatures that form *Utricularia*'s

Aechmea fasciata

36. Many plants, such as *Aechmea fasciata*, shown here, form tightly packed vases with their leaves, and these retain water. They also trap small animals and insects; and the leaves are capable of absorbing dissolved minerals from the water.

Pink blossom

prey must live there too, and so the reservoirs of water must be far from temporary. Many of these vaselike plants are found in North and South America, most perhaps in the vast floral territory of Brazil. There are altogether about sixty genera and fourteen hundred species, but by no means all of them form vases.

Many of these plants prefer shady situations, and that perhaps is why they are so prolific in Brazil, where foliage is often extremely dense. Some do grow well in the sun, however, and have taken well in Australia and New Zealand. They can be found growing on fallen logs, on cacti, or high up on open mountain plateaus.

Some of them are very colorful, too, especially the genus Vriesia. One has wine-red stripes on green leaves with yellow-and-red flowers[75]. Another has light-green leaves with red-and-yellow flowers[72]. Many others are brightly colored, and almost all can be raised at home. *Aechmea* has some very attractive species, and they will grow very large, up to two feet across. Others, such as *Guzmania, Neorgelias, Billbergia,* and *Nidularium,* ranging from Mexico to Chile, all have the rosette vases, and many of these can be obtained from nurseries.

These plants are discussed in Chapter 8, which describes how to grow them at home, where they can be observed daily by anyone with the time and interest to do this.

8. Growing Carnivorous Plants at Home

$\mathcal{A}$lthough it may seem that the home raising of such highly specialized plants under artificial conditions would be difficult, this is not always so. It is true that some of them require special soil preparation, well-balanced moisture care, and certain levels of temperature and humidity, but many of them can be grown in the open in temperate climates, and in greenhouses even up north near the Arctic Circle. There are many places where plants, tubers, or seeds can be obtained from good horticultural dealers, and once these are established at home they can be multiplied with a little care and patience.

Undoubtedly the Venus's-flytrap captures the imagination of most people who want to observe carnivorous plants; it is such a master-piece of mechanical precision, and it can be fed with small flies caught by hand. This one will be discussed first, but some of the others can prove equally interesting and quite trouble-free.

Venus's-Flytrap. *Dionaea muscipula* can be obtained in the form of plants[23] or dormant tubers, usually mere bulbils (small bulbs broken off from the sides of mature ones), which can be raised to healthy plants with very little trouble. Seeds can also be obtained, but naturally reach maturity and large plant size in a longer time. Both the bulbils and the seeds should be planted in pots half full of broken earthenware crockery topped with a compost of washed silver sand mixed with loam, plus chopped-up swamp moss, such as sphagnum, and fern fiber that has been cleaned of dust. The silver sand should be a mere 10 percent in sifted loam, if possible, because the whole mixture must always be wet. The pots should then be immersed to the top in damp sphagnum, with some small live cuttings planted in places on the surface of the latter.

It will have become obvious that sphagnum moss is favored for growing carnivorous plants at home, and that no other is suggested. There is a very good reason for this. Not only is sphagnum the plant life among which many of these carnivorous plants grow in the wild state, but it also reduces the level of nitrogen in the soil and water and makes it more acid. This increases a plant's need for animal food, and so ensures a healthy trapping condition.

These plants must be kept moist with rainwater or soft water from some other source, never tap water or salt water; and once the plant shows above ground, it must never have water actually splashed on it, because water will turn the leaves brown and they will deteriorate. These plants can be kept in a cool but frost-free place, and they will then grow well and flower in the summer. The flowers are attractive, but they should be clipped off early in their development until the plant has grown into a large well-formed rosette, after which they can be encouraged to mature fully for seed production if this is desired.

If no flies are around for the plant to attract, or none are available, the traps can occasionally be fed tiny pieces of lean raw beef, and they will then be seen to operate just as if insects had alighted on them; but it could prove unwise to feed more than one or two leaves at a time. The most important precautions to be taken with these plants are to ensure that the water used for them is free from minerals, especially from salt, and that it does not touch the leaves, even though rain does touch them in the wild state. In a city, however, water is not quite so clean; it may contain sulfur and other contaminating materials.

Sundews. The *Drosera* species are probably the most interesting plants to raise at home, and these can be kept in pots too. In fact, the instructions offered for the planting of *Dionaea* apply to *Drosera* too, except that the pots in which they are planted need to be placed in saucers of water, unless there is a cool moist greenhouse environment available. If saucers of water are used, a glass or plastic cover can be mounted two or three inches above each plant to help in producing a humid atmosphere.

These plants can also be grown from seed or by a division of rhizomes. There is less of a problem about watering them than there is with *Dionaea*. The water must be soft for *Drosera,* too, but it likes a fine spray falling from above every now and then. Two of the most attractive species to raise when they can be obtained are *Drosera capensis,* which has rosettes of purplish leaves and reddish-purple flowers, and *Drosera cistiflora* from South Africa, which can have flowers ranging in color from white to scarlet.

Pitfall Traps. Of all the pitcher-type plants, *Sarracenia* may be the easiest to rear. It is not so easy to observe and inspect perhaps as *Nepenthes,* but the latter requires much more room to grow. Since there are ten species of *Sarracenia* in North America, they are not difficult to obtain. *Sarracenia purpurea* (see Figure 14 (top), and 15) is a handsome species that can be grown either in a greenhouse or beside a pond in a bog garden. The pitcher of this species has a wide body and is relatively short compared with so many of the other species of *Sarracenia,* so its habits are easier to observe.

Other interesting species, some of which have colorful markings, are *Sarracenia mitchelliana* and *Sarracenia leucophylla,* but practically all species can be raised at home if a cool, moist greenhouse is available. They need much the same moist, boggy conditions as do *Drocera* and *Dionaea.*

Nepenthes can also be grown at home if enough space is available, and it will not cease to fascinate the grower, for its development is as intriguing as are its carnivorous habits, perhaps more so. A great many have been grown in Europe, and growers there have produced some colorful hybrids. They must, however, be kept in a warm, moist atmosphere, with the temperature not below 65°F in winter, and kept at 70° to 80°F in summer.

The soil mixture *Nepenthes* likes best is made up of peat moss, sphagnum, and some charcoal lumps, and it must be moist and boggy. All these carnivorous plants are fussy about their water. It must always be soft, or be rainwater.

Butterworts. *Pinguicula* can be successfully grown, under much the same conditions as those required by *Nepenthes. Pin-*

guicula must have soft water; minerals in the water will kill it. It also must be kept constantly moist, which is best done by standing its pot in water and spraying it now and then with soft water. The pot or pan in which it is grown must have plenty of broken earthenware crockery and a loose mixture of loam, coarse sand, fibrous peat, and sphagnum.

Bladderworts. These underwater traps can be cultivated in ponds, but anyone interested in their habits must make an effort to raise them in an aquarium that has been stocked with a few daphnia and cyclops as prey for the bladders. The plants are best grown against the front glass of the aquarium, so that one can get close enough to them to use a high-powered magnifier for observation. Species of *Utricularia* in the United States can be grown from seeds, which lie on the bottom of ponds. Some species grow on land on the banks of swamps and other wet places, and these are not difficult to raise in pans or baskets of peat and sphagnum. The containers must be kept in water to keep the soil constantly moist, and so that the bladders can be under water to catch prey.

Some need to be in a warm greenhouse, and those that have water-retaining tubers can be kept outside pools or aquariums so long as they are never allowed to dry out. Examples of these are *U. endressii* and *U. alpina.* Of those that should be kept in water-immersed pans, *U. longifolia* and *U. humboldtii* might be the easiest to obtain. But they all come from South America, and so may have to be obtained from dealers specializing in exotic plants.

The buds of *Aldrovanda,* which sink to the bottom of the water in winter, can, when found or purchased, be raised in lily ponds at home, but because they must have shade they must be grown among reeds or water lilies in areas exposed to heat. For this reason, and because they remain submerged except when flowering, they are not easily observed, so few people bother to rear them.

Bromeliads. Although most other carnivorous plants are rather difficult to grow, many of the vase traps are good greenhouse plants, or even, in suitable climates, easily grown in a garden. These include some of the *Aechmea* and *Vriesia* species, which are also among the most colorful. Some bromeliads grow on trees, some on rocks, but many do well in soil, and it is these last that are the pride of the home gardener; and anyone wishing to study the possible trapping habits of vases should choose these. They will grow quite easily in a greenhouse, but there are usually no animals there, so a more exposed location is needed. A warm climate is therefore necessary, and perhaps a greenhouse just for the winter.

Aechmea discolor has a beautiful vase, and so has *Aechmea fasciata*. Then there are *Vriesia carinata, V. fenestralis, V. splendens,* and *V. hieroglyphica,* all equally beautiful. Both *Aechmea* and *Vriesia* like shady, humid locations, for they come from South American jungles and do not take kindly to sunshine or dry air. Again, rainwater or soft water is essential, and plenty of it. They can be kept in large pots packed with osmunda fiber, which is obtained from a few species of fern.

Species of *Guzmania, Billbergia, Nidularium,* and *Neorgelia* add many more colorful vases to the list that can be purchased and grown successfully at home, but these need fairly high temperatures and plenty of water sprayed onto their foliage as well fed to their roots. There are a few from the mountain heights of South America that are hardy in cool climates, but most of the picturesque ones, and the large ones capable of trapping fair-sized animal life, are tropical.

Bibliography

Darwin, Charles, *Insectivorous Plants*. London, John Murray, 1908.

Duddington, C. L., *The Friendly Fungi*. London, Faber & Faber, 1957.

Erickson, Rica, *Plants of Prey in Australia*. Osborne Park, Western Australia, Lamb Publications, 1968.

Garrett, S. D., *Soil Fungi and Soil Fertility*. Oxford, Pergamon Press, 1963.

Graf, Alfred Byrd, *Exotica—Pictorial Cyclopedia of Exotic Plants*. Rutherford, New Jersey, Roehrs Company, 1968.

Lloyd, Francis Ernest, *The Carnivorous Plants*. Waltham, Massachusetts, Chronica Botanica Co., 1942.

Glossary of Some Terms Used in This Book

algae   A simple form of plant life that includes both minute and giant forms from the simple green scum on still water to giant seaweeds and kelp.

animalcules   Literally, animals too small to be seen by the naked eye.

asexual   Without sex or sexual function.

catalyst   Any substance that changes the speed of a chemical reaction without itself being changed.

coenocytic   Common cells.

conidium (pl. conidia)   An asexual spore in certain fungi.

desmids   A group of microscopic freshwater algae differing from diatoms in their green color and the absence of a siliceous covering.

diatoms   A group of microscopic algae with siliceous coverings found in large numbers on the seafloor, which multiply by division or conjugation.

endozoic   Living inside or passing through an animal, as when a seed is swallowed by an animal.

enzyme   A complex protein, created by a living organizm, that brings about chemical changes.

euglena   A single-celled animal with a "flagellum" or tail capable of agitating the water enough to produce locomotion.

fungi   Simple plant organisms having no chlorophyll and living on organic material.

conjugation   Cellular contact between siliceous protozoans.

haustorium   A "sucker" or invading process of a parasitic plant.

hydrodynamic   Relating to the force created by water.

hydrolysis   The chemical decomposition of substances through the fixation of the elements of water in compounds.

· 83 ·

hydroponic   Pertaining to the cultivation of plants in water containing chemicals but no soil, or soil only as a support.

hypha   A threadlike segment of the body of a fungus.

hypotonic   Lower in tone or tension.

infusiorian   Any number of minute protozoa found in decaying organic matter.

isotonic   Having equal tensions.

mucilage   A gumlike or viscous substance in plants or a lubrication substance in animal bodies.

mycelium   The body of a fungus.

nematodes   Parasitic threadlike or round worms.

parasite   An organism that lives in or on another living organism.

pedicel   The stalk supporting a single flower.

peptone   A soluble digest of protein created by the action of pepsin, a protease.

pollination   The fertilization of plants by pollen.

predaceous   Living on prey.

protease   An organic compound that breaks down proteins.

protozoa   The lowest division of the animal kingdom; single-celled or multi-celled organisms of a single kind of tissue.

rhizoid   A hairlike organ for anchoring, a rootlike process.

rhizopod   An animalcule with pseudopodia for locomotion, or the mycelium of a fungus.

saprophyte   An organism living on dead or decaying organic matter.

septum (pl. septa)   A dividing wall or membrane in soft tissue, as in a fungus.

sessile   Attached at the base; without a stalk.

siliceous   Containing silica.

sphagnum   Bog or peat mosses.

trichome   An outgrowth, such as a hair, filament, or thread.

trophic   Relating to nutrition.

tubercle   A small prominence, a small tuber, or a warty growth.

unicellular   Consisting of a single cell.

vascular   A system of vessels or ducts for carrying blood or sap.

Glossary
of Scientific Terms

1. *Acaulopage* spp.
2. *Aechmea discolor*
3. *Aechmea fasciata*
4. *Aldrovanda vesiculosa*
5. *Amphiprion percula*
6. *Arthrobotrys cladodes*
7. *Arthrobotrys oligospora*
8. *Bdellospora helicoides*
9. *Billbergia*
10. *Biovularia*
11. *Bromeliaceae*
12. *Byblis gigantea*
13. *Byblis linifolia*
14. *Cephalotus follicularis*
15. *Cochlonema dolichosporum*
16. *Cochlonema verrucosum*
17. *Cordyceps* spp.
18. *Dactylella bembicoides*
19. *Dactylella cyanopaga*
20. *Dactylella ellipsosporum*
21. *Dactylella tylopaga*
22. *Darlingtonia californica*
23. *Dionaea muscipula*
24. *Drosera capensis*
25. *Drosera cystiflora*
26. *Drosera intermedia*
27. *Drosera leucoblasta*
28. *Drosera lovella*
29. *Drosera occidentalis*
30. *Drosera peltata*
31. *Drosera pulchella*
32. *Drosera pycnoblasta*
33. *Drosera pygmaea*
34. *Drosera rotundifolia*
35. *Drosophyllum lusitanicum*
36. *Endocochlus asteroides*
37. *Genlisea repens*
38. *Guzmania*
39. *Harposporum anguillulae*
40. *Heliamphora nutans*
41. *Heliamphora tatei*
42. *Neorgelias*
43. *Nepenthes khasiana*
44. *Nepenthes maxima*
45. *Nepenthes rafflesiana*
46. *Nepenthes rajah*
47. *Nidularium*
48. *Paphiopedilum*

49. *Pedileps dactylopaga*
50. *Pedilospora dactylopaga*
51. *Pinguicula alpina*
52. *Pinguicula lutea*
53. *Pinguicula macrophylla*
54. *Pinguicula vulgaris*
55. *Polypompholyx* spp.
56. *Roridula dentata*
57. *Sarracenia flava*
58. *Sarracenia leucophylla*
59. *Sarracenia mitchelliana*
60. *Sarracenia psittacina*
61. *Sarracenia purpurea*
62. *Sommerstorffia spinoza*

63. *Stylopage hadra*
64. *Trichothecium cytosporum*
65. *Utricularia alpina*
66. *Utricularia dichotoma*
67. *Utricularia endressi*
68. *Utricularia humboldtii*
69. *Utricularia longifolia*
70. *Utricularia purpurea*
71. *Utricularia vulgaris*
72. *Vriesia carinata*
73. *Vriesia fenestralis*
74. *Vriesia hieroglyphica*
75. *Vriesia splendens*

Index